Math Is Everywhere!
MATH IN THE GARDEN

By Kieran Shah

Gareth Stevens
PUBLISHING

Please visit our website, www.garethstevens.com. For a free color catalog of all our high-quality books, call toll free 1-800-542-2595 or fax 1-877-542-2596.

Library of Congress Cataloging-in-Publication Data

Names: Shah, Kieran.
Title: Math in the garden / Kieran Shah.
Description: New York : Gareth Stevens Publishing, [2017] | Series: Math is everywhere! | Includes index.
Identifiers: LCCN 2016027685| ISBN 9781482454413 (pbk. book) | ISBN 9781482454420 (6 pack) | ISBN 9781482454437 (library bound book)
Subjects: LCSH: Arithmetic–Juvenile literature. | Gardening–Juvenile literature.
Classification: LCC QA115 .S4974 2017 | DDC 513–dc23
LC record available at https://lccn.loc.gov/2016027685

First Edition

Published in 2017 by
Gareth Stevens Publishing
111 East 14th Street, Suite 349
New York, NY 10003

Copyright © 2017 Gareth Stevens Publishing

Editor: Therese Shea
Designer: Sarah Liddell

Photo credits: Cover, p. 1 Alexander Raths/Shutterstock.com; pp. 5, 24 (garden) Verena Matthew/Shutterstock.com; p. 7 Le Do/Shutterstock.com; p. 9 Darko Vrcan/Shutterstock.com; p. 11 Kitzero/Shutterstock.com; pp. 11, 24 (seeds) Billion Photos/Shutterstock.com; pp. 13, 24 (seedling) kirillov alexey/Shutterstock.com; p. 15 INSAGO/Shutterstock.com; p. 17 AndrisA/Shutterstock.com; p. 19 Sofiaworld/Shutterstock.com; p. 21 Ian 2010/Shutterstock.com; p. 23 belushi/Shutterstock.com.

All rights reserved. No part of this book may be reproduced in any form without permission in writing from the publisher, except by a reviewer.

Printed in China

CPSIA compliance information: Batch #CW17GS: For further information contact Gareth Stevens, New York, New York at 1-800-542-2595.

Contents

Our Garden. 4

Seedlings. 10

Time to Plant 14

A Gift 20

Words to Know 24

Index. 24

We have a garden.
It's in front of our house.

A bed is a part
of a garden.
This bed has
many colors.

This bed has red flowers.

Seedlings are baby plants.
They grew from seeds.

seeds

The seedling on the right is taller.

Dad plants 3 flowers.
I plant 1 more.
We plant 4 flowers in all.

15

My sister Kate plants
5 flowers.
I plant 1 more.
We plant 6 flowers in all.

17

I water the plants.
It's my job!

I pick 5 flowers.
I pick 5 more.
I pick 10 flowers in all.

The flowers are
for Grandma!

Words to Know

 garden

 seedling

 seeds

Index

bed 6, 8
flowers 8, 14, 16,
    20, 22
seedlings 10, 12
seeds 10

9